Learning About Landforms

Islands

Ellen Labrecque

Heinemann
LIBRARY

Chicago, Illinois

Edited by Rebecca Rissman, Daniel Nunn, and Catherine Veitch
Designed by Steve Mead
Picture research by Elizabeth Alexander
Production by Victoria Fitzgerald
Originated by Capstone Global Library
Printed and bound in China

17 16 15 14 13
10 9 8 7 6 5 4 3 2 1

Library of Congress Cataloging-in-Publication Data
Labrecque, Ellen.
Islands / Ellen Labrecque.
pages cm.—(Learning about landforms)
Includes bibliographical references and index.
ISBN 978-1-4329-9534-8 (hb)—ISBN 978-1-4329-9540-9 (pb) 1.
Islands—Juvenile literature. I. Title.
GB471.L33 2014
551.42—dc23   2013014818

Acknowledgments
We would like to thank the following for permission to reproduce
photographs: Alamy pp. 8 (© Travel Pictures), 20 (© Emmanuel
LATTES); Getty Images pp. 9 (Torsten Blackwood/AFP), 11 (Michael
Dunning/Photographer's Choice), 13 (PARK YEONG-CHEOL/AFP),
14 (Bob Stefko/Photodisc), 15 (Viviane Ponti/Lonely Planet Images),
17 (Patrick AVENTURIER/GAMMA), 18 (© Laszlo Podor/Flickr),
19 (Popperfoto); naturepl.com pp. 22 (© Doug Allan), 29 (© Aflo);
Science Photo Library p. 10 (Christian Darkin); Shutterstock pp.
4 (© photobank.kiev.ua), 5 (© Tatiana Popova), 12 (© Igor
Plotnikov), 16 (© SUPACHART), 21 (© Konstantin Stepanenko),
23 (© Steffen Foerster), 24 (© Pierre Leclerc), 25 (© Subbotina
Anna), 26 (© Deborah Kolb), 27 (© Aleksandar Todorovic),
28 (© Laborant); SuperStock p. 6 (Planet Observer/Universal
Images Group).

Cover photograph of Parata Headland and Sanguinaires Islands
reproduced with permission of Corbis (© Marc Dozier/Hemis).

Every effort has been made to contact copyright holders of material
reproduced in this book. Any omissions will be rectified in subsequent
printings if notice is given to the publisher.

All the Internet addresses (URLs) given in this book were valid at
the time of going to press. However, due to the dynamic nature of
the Internet, some addresses may have changed, or sites may have
changed or ceased to exist since publication. While the author
and publisher regret any inconvenience this may cause readers, no
responsibility for any such changes can be accepted by either the
author or the publisher.

Contents

Some words are shown in bold, **like this.** You can find out what they mean by looking in the glossary.

What Are Landforms?

Look around when you are outside. You may see mountains, hills, valleys, caves, or islands. Earth is made of these natural landforms. This book is about islands.

This island is in the Maldives.

Earth is always changing. Earth's landforms are made in different ways over millions of years. Some islands are made from volcanoes. Others are made when large areas of land split and move over time.

What Are Islands?

Islands are land surrounded on all sides by water. Islands are found in oceans, rivers, lakes, or ponds. Islands are different sizes. Greenland is the world's largest island (about 840,000 square miles).

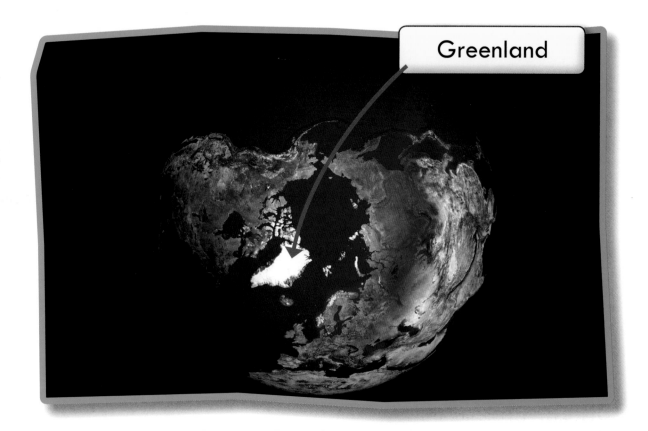

Greenland

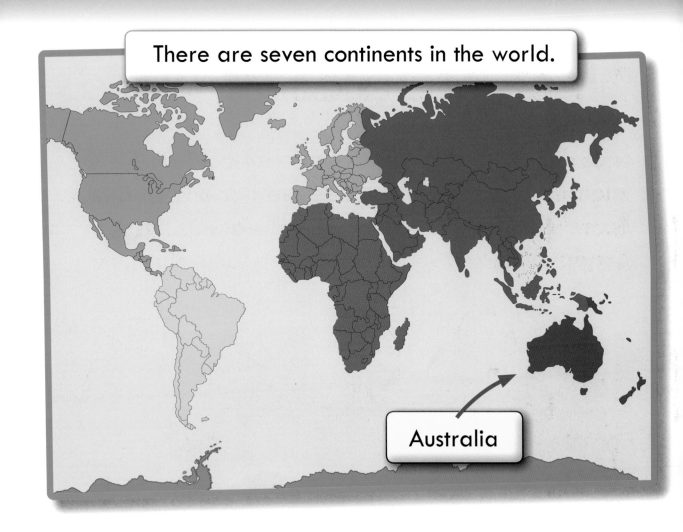

There are seven continents in the world.

Australia

Australia is surrounded on all sides by water, but Australia is not an island. Australia is too big to be an island (about 3 million square miles). Australia is a **continent**. Continents are much bigger than islands.

Different Types of Islands

Islands can be different. Some are cold and covered in ice. Other islands are hot and **humid**. Some islands have lots of animals and people. Others do not have many.

This island is heart-shaped. It is called Lovers' Island.

Manukan Island, Malaysia

Islands are made in different ways. There are six types of islands. These are continental, tidal, barrier, coral, volcanic, and artificial.

Continental Islands

Continental islands were once part of a **continent**. Scientists think that millions of years ago, there was only one large continent on Earth. Over time, this continent broke into pieces. Some of these pieces became continental islands.

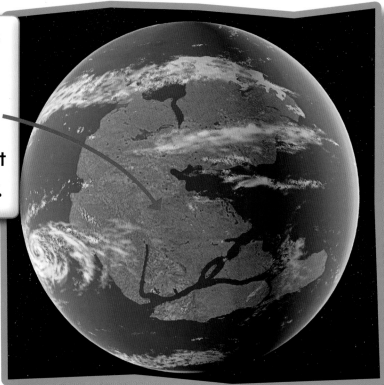

All the countries on Earth were once joined together in one massive continent called Pangaea.

Great Britain is a continental island.

Continental islands are also made when the sea floods over land and cuts off other parts of land. The British Isles were made into an island in this way. The British Isles were once part of Europe.

Tidal Islands

Tidal islands are types of continental islands. They are still joined by land to the **mainland,** but the connection is underwater most of the time.

Mont Saint-Michel is a tidal island in France.

path

A few times a year, Jindo and Modo are joined to one another for one hour.

Jindo and Modo are two tidal islands near South Korea. A few times a year, the sea level goes down and a path opens up between the two islands.

Barrier Islands

Barrier islands are made when the sea piles up sand into **sandbars**. The sandbars rise above the seawater and make islands. Barrier islands are next to coastlines. They protect the coastline from storms, waves, and wind.

Padre Island in Texas is the world's longest barrier island.

These barrier islands are off the coast of Brazil.

Many barrier islands were also made during the most recent **ice age** (11,700 years ago).

Coral Islands

Coral islands are made of billions of tiny **exoskeletons**, or dead animals' bones. Other materials, like rock and sand, mix with the coral. This helps the island grow big and rise from the bottom of the sea.

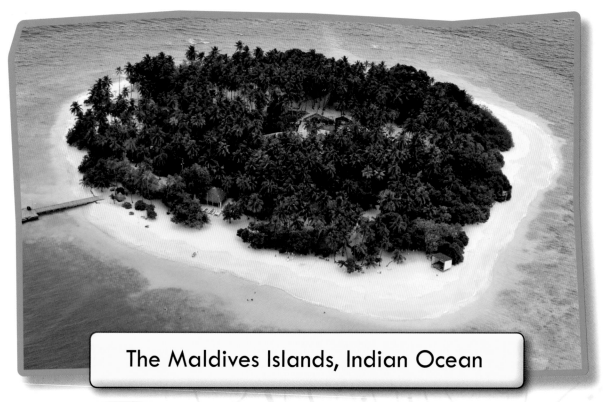

The Maldives Islands, Indian Ocean

Tuvalu Islands, South Pacific

The storms that create coral islands can also damage and destroy them. Waves attack one side and dump material on the other.

Volcanic Islands

When an underwater volcano **erupts, lava** flows out and down the sides of the volcano. The lava cools and piles up. When the volcano grows big enough, it rises above the water and forms an island.

The Hawaiian Islands are volcanic islands.

Scientists study the new plant and animal life on Surtsey Island.

In 1963, Surtsey Island was created when a volcanic eruption spewed hot lava into the Atlantic Ocean near Iceland.

Artificial Islands

Artificial islands are another type of island. They are not really landforms, because people created them. Hundreds of years ago on Lake Titicaca, the Uros people made islands out of **reeds**, off the coast of Peru.

Homes and boats on Lake Titicaca are also made from reeds.

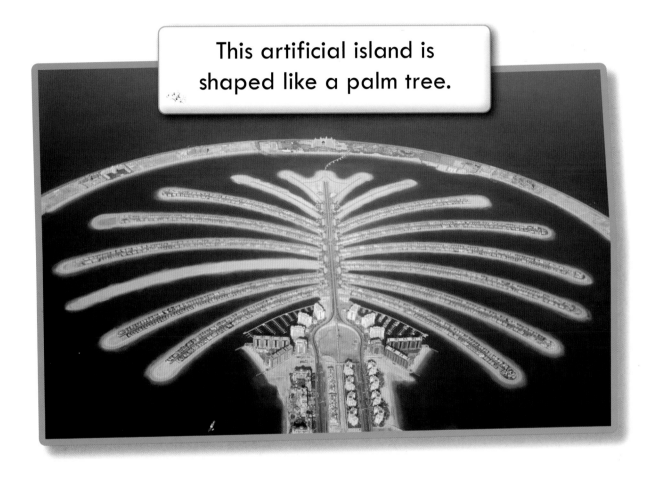

This artificial island is shaped like a palm tree.

In Dubai in the United Arab Emirates, people have created islands shaped like palm trees and a map of the world. The islands are made out of sand. The sand has been moved from the Persian Gulf and sprayed into shapes near the shore.

Island Animals

Different animals live on islands. The animals that live on continental islands are similar to the creatures that live on the nearby **mainland.**

This walrus lives in Greenland.

Creatures swim or fly to volcanic islands from all over the world. Animals are also carried to volcanic islands on floating plants that have been swept out to sea.

This blue-footed booby lives on the Galapagos Islands.

Plants and Trees

Some islands are home to beautiful plants. Some plant seeds arrived on islands by drifting in the ocean. Coconut tree seeds are enclosed in hard shells that can float a long time. The seeds of red mangrove trees often float to new places, too, along a coastline.

bird of paradise
flower, Hawaii

Seeds can also travel to islands by wind, and birds can carry seeds to islands. The seeds get stuck on the birds' feet or feathers, or they get tucked in their droppings!

Island People

Many scientists say the first people to live on the islands in the Pacific Ocean came from Southeast Asia. These people sailed in canoes across the sea around 3,000 or 4,000 years ago. They were called Polynesians.

These Polynesians live on an island in the Pacific Ocean.

Java, Indonesia

Today, millions of people live on islands. Java in Indonesia is the world's most crowded island. More than 100 million people live there.

Islands Today

Many people love islands, especially **tropical** or warm ones. People love to swim, dive, and play on island beaches.

But islands are also at risk from rising sea levels and fierce weather such as **tsunamis** and hurricanes. Scientists study ways of saving islands. They hope to make a difference, so that we can always enjoy islands.

Glossary

continent one of seven of the largest landmasses on Earth

erupt explode or break open

exoskeleton hard outer shell that protects some animals

humid hot and sticky

ice age period of time when large sheets of ice covered large areas of land

lava hot, melted rock that comes out of a volcano

mainland big area of land

reeds tall, stiff grass that grows on or near water

sandbar exposed or underwater ridge of sand built by waves offshore

tropical weather that is hot and sticky

tsunami destructive sea wave caused by an underwater earthquake

Find Out More

Books to read

Ganeri, Anita. *Ocean Divers* (Landform Adventurers). Chicago: Raintree, 2012.

Green, Jen. *Islands Around the World* (Geography Now!). New York: PowerKids, 2009.

Penrose, Jane. *Landforms* (Investigate Geography). Chicago: Heinemann Library, 2010.

Web sites to visit

Facthound offers a safe, fun way to find web sites related to this book. All the sites on Facthound have been researched by our staff.

Here's all you do:
Visit **www.facthound.com**
Type in this code: 9781432995348

Index